U0240953

不可思议的恐龙知识

Incredible Dinosaur Facts

[英] 露丝·欧文/著

刘颖/译

汉英对照
恐龙科普

江苏凤凰美术出版社

全家阅读
小贴士

★ 每天空出大约10分钟来阅读。

★ 找个安静的地方坐下，集中注意力。关掉电视、音乐和手机。

★ 鼓励孩子们自己拿书和翻页。

★ 开始阅读前，先一起看看书里的图画，说说你们看到了什么。

★ 如果遇到不认识的单词，先问问孩子们首字母如何发音，再带着他们读完整句话。

★ 很多时候，通过首字母发音并听完整句话，孩子们就能猜出单词的意思。书里的图画也能起到提示的作用。

最重要的是，感受一起阅读的乐趣吧！

扫码听本书英文

Tips for Reading Together

- Set aside about 10 minutes each day for reading.

- Find a quiet place to sit with no distractions. Turn off the TV, music and screens.

- Encourage the child to hold the book and turn the pages.

- Before reading begins, look at the pictures together and talk about what you see.

- If the child gets stuck on a word, ask them what sound the first letter makes. Then, you read to the end of the sentence.

- Often by knowing the first sound and hearing the rest of the sentence, the child will be able to figure out the unknown word. Looking at the pictures can help, too.

Above all enjoy the time together and make reading fun!

Contents 目录

神奇的恐龙
Amazing Dinosaurs

恐龙是有史以来最神奇的动物之一。
它们在地球上生存了1.7亿多年！

Dinosaurs were some of the most amazing animals that ever lived.
They lived on Earth for over 170 million years!

棘龙 Spinosaurus
(SPY-no-SAW-rus)

科学家已发现了700多种不同恐龙的化石。

Scientists have found the **fossils** of 700 different kinds of dinosaurs.

化石 fossil

三角龙
Triceratops
(try-SERA-tops)

恐龙英文单词的字面意思是"可怕的蜥蜴"。
The word dinosaur means "terrible lizard".

最大的和最小的
Biggest and Smallest

巴塔哥巨龙是最大的
恐龙之一。
它有5层楼那么高。

Patagotitan was one of the biggest dinosaurs.
It was as tall as a five-storey building.

巴塔哥巨龙
Patagotitan (patter-go-TIE-tun)

小盗龙是最小的恐龙之一。
它甚至比鸡还小。

Microraptor was one of the
smallest dinosaurs.
It was smaller than a chicken.

阿根廷龙 Argentinosaurus
(ar-jen-TEE-no-SAW-rus)

小盗龙 Microraptor
(MY-crow-rap-ter)

阿根廷龙的体重相当于18头大象的重量!
Argentinosaurus was as heavy as 18 elephants!

最多的角
The Most Horns

三角龙有3个角，但华丽角龙有15个角！

Triceratops had three horns – but Kosmoceratops had 15 horns!

华丽角龙
Kosmoceratops
(kos-mo-SERA-tops)

华丽角龙的头上有2个长长的角。

它的脸上长着3个角。

Kosmoceratops had two long
horns on its head.
It had three horns on its face.

短角 **short horns**

鼻角 **nose horn**

面颊角 **cheek horn**

它的头骨后方还有10个短短的角。

It had ten short horns behind its skull.

会奏音乐的恐龙
A Musical Dinosaur

副栉龙的头顶有个长长的冠饰。

冠饰内部有空心管。

Parasaurolophus had a long **crest** on its head.

The crest had hollow tubes inside.

冠饰 **crest**

科学家认为副栉龙能将空气吸入空心管里。
这可能会产生响亮的鸣叫声，就像管风琴的
声音一样。

Scientists think this dinosaur could breathe air into the tubes.
This might have made a loud, booming noise – like an **organ**.

副栉龙 **Parasaurolophus
(para-saw-RO-lo-fuss)**

坚硬的头！
Hard-Headed!

厚头龙生来就长着"头盔"。

Pachycephalosaurus was a dinosaur with a built-in helmet.

厚头龙 **Pachycephalosaurus**
(pak-ee-SEFF-al-oh-SAW-russ)

它的头骨厚达25厘米。
它将自己的头当作撞锤用！

Its skull was 25 cm thick.
It used its head as a battering ram!

裂纹
cracks

头骨化石有裂纹。
这说明这头恐龙用头撞过东西。
Fossil skulls have cracks.
This shows that the dinosaurs butted heads.

最长的爪子
The Longest Claws

镰刀龙的手上长着3根长长的爪子。
但它并不用巨爪来猎食其他恐龙。

Therizinosaurus had three very
long claws on its hands.
But it didn't use its claws to
hunt other dinosaurs.

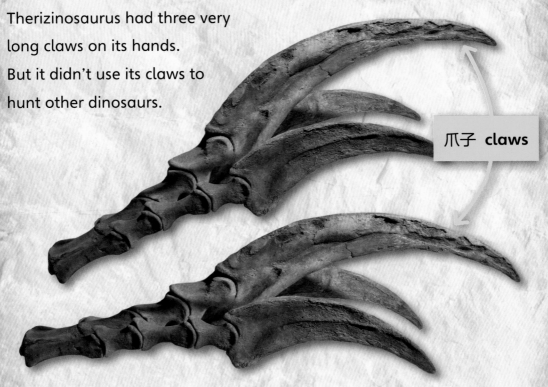

爪子 claws

它是食草动物。
It was a plant-eater.

镰刀龙 **Therizinosaurus**
(ther-ih-ZEE-no-SAW-rus)

镰刀龙的爪可以长到1米长。
The claws could grow to 1 metre long.

最强的咬合力
Most Powerful Bite

哪种恐龙时代的物种拥有最有力的下颚？

科学家发现短吻鳄能咬断马腿。

Which species had the most powerful jaws in Age of Dinosaurs?

Scientists found an alligator could snap the leg off a horse.

短吻鳄
alligator

接着，他们发现霸王龙的牙齿能咬碎汽车！

Then they found out that a Tyrannosaurus rex could crush a car with its teeth!

霸王龙 Tyrannosaurus rex
(tie-RAN-oh-SAW-rus rex)

大脑的大小很重要
Brain Size Matters

伤齿龙是一种小型恐龙。

它的大脑和李子一样大。

Troodon was a small dinosaur.

Its brain was as big as a plum.

伤齿龙 Troodon
(TRO-uh-don)

它非常聪明。

It was very smart.

剑龙体形巨大。

它的大脑也和李子一样大。

它不太聪明。

Stegosaurus was a big dinosaur.

Its brain was also as big as a plum.

It was not very smart.

剑龙
Stegosaurus
(STEG-oh-SAW-rus)

恐龙蛋
Dinosaur Eggs

所有恐龙都是卵生动物。

大型恐龙的蛋和足球一样大。

All dinosaurs hatched from eggs.

The eggs of big dinosaurs were the size of a football.

蛋壳 **eggshell**

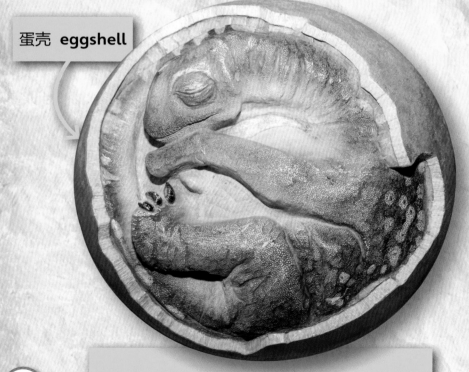

恐龙胚胎模型
a model of a baby dinosaur inside its egg

恐龙蛋 dinosaur eggs

最小的恐龙蛋只有2厘米长。
The tiniest dinosaur eggs are 2 cm long.

阿根廷龙宝宝
baby Argentinosaurus

词汇表 Glossary

冠饰 crest

动物头部由骨头、羽毛或皮肤构成的区域。

An area made of bone, feathers or skin on an animal's head.

化石 fossil

存留在岩石中几百万年前的动物和植物的遗体。

The rocky remains of an animal or plant that lived millions of years ago.

管风琴　organ

类似于钢琴的一种大型
乐器，有琴键和音管。

A large musical instrument, like a
piano, with keyboards and pipes.

科学家　scientist

研究自然和世界
的人。

A person who studies
nature and the world.

恐龙小测验 Dinosaur Quiz

① 科学家发现了多少种恐龙？

How many different kinds of dinosaurs have scientists found?

② 巴塔哥巨龙比小盗龙大还是小？

Was Patagotitan bigger or smaller than Microraptor?

③ 三角龙和华丽角龙，哪种恐龙有更多的角？

Which had more horns – Triceratops or Kosmoceratops?

④ 霸王龙的咬合力有多强？

How strong was T. rex's bite?

⑤ 你认为哪个恐龙知识最不可思议？

Which dinosaur fact do you think is the most amazing?